Bobo Learns About Heat

For my inquisitive boys who love science,
just a little less than they love ice cream - p.k.

For those kids who are always asking questions
about the world around us - r.c.

Printed in the United States of America

First Printing, 2017

ISBN 978-0-9972441-1-3

www.bobbojabba.com

Bobo Learns About Heat

A Bobo & Jabba story

Written by
Praveen Krishnan

Illustrated by
Rajeev Char

One hot summer day, Bobo and his best friend Jabba were playing in the woods...

Come on, one more game, Jabba, PUHLEEZ?
I'm tired, Bobo. Let's go.

Bobo, I think I hear the ice cream bear.

Do you want to get ice cream...? Hey, wait...

Bobo and Jabba choose all their favorite flavors of ice cream.

They find a nice spot in the woods to sit and enjoy their ice cream.

Why does ice cream melt?
YUM!

JABBA!! Why does ice cream MELT?
OK! OK! I'll tell you!

HEAT melts the ice cream.

What's HEAT, Jabba?

NO! Tell me! Tell me!
Can I eat my ice cream first??

OK, FINE!
I'll tell you.
Yay!!

The Sun sends Heat
and Heat melts the ice cream.

But Jabba! The sun is over there and my ice cream is over here.

How can the Sun make my ice cream melt??

The Sun sends out a lot of energy called HEAT.

Now, let's get my favorite tool to see tiny things... my **microscope**.

Look inside my ice cream with our super microscope.

Do you see tiny blobs inside the Ice Cream? They are called **Atoms**.

Yes, Atoms!! I remember.

Ice cream is cold and the atoms inside cold ice cream do not move much.

When ice cream is in the sun,
something very interesting happens....

The **heat** from the sun gets all the **atoms** to **move** and **dance**...

When they **move around** a lot, the ice cream loses its shape and **melts**!

AH HA!
I get it.
Good!

So, the sun sends heat to the ice cream
Yes!

And heat shakes the atoms, pushing them away from each other, making ice cream spread out and MELT!!
You GOT IT, Bobo!

Finally... Now I can eat my YUMMMMY ice cream!

HEY!! Who took my ice cream?!

HEAT!

THE END.

Authors' Note

The goal of this book has been to convey the science behind heat and how it transfers to things like ice cream.

We wanted a fun way to talk about heat and what is more fun than ice cream!

Molecules and atomic structures are only observable using powerful electron microscopes. Jabba's super duper microscope is a generation more advanced than any of those (we think).

Radiation is the method by which energy from the sun arrives on Earth.

We didn't think Bobo would mind if we kept things simple.

Thank you,

Peek & Char

Questions for Kids

1. What are some other things that melt when the sun shines?
2. What melts faster: ice or butter?
3. Can you keep ice cream from melting in the sun?
4. How many pants did Jabba change throughout the story?
5. What happens when you leave ice cream out in the sun all day?
6. What is the best way to eat ice cream from a cone?

Made in United States
North Haven, CT
05 November 2026

21765289R00022